The World's COOLEST Jobs

ANIMAL HANDLER

Alix Wood

PowerKiDS
press

New York

Published in 2014 by The Rosen Publishing Group, Inc.
29 East 21st Street, New York, NY 10010

Editor for Alix Wood Books: Eloise Macgregor
Designer: Alix Wood
US Editor: Joshua Shadowens
Researcher: Kevin Wood

Photo Credits: Cover, 5, 6, 7 bottom, 8, 9, 10, 12, 13, 14, 15 bottom, 16,
17, 18, 19, 21, 23, 26, 27, 28, 29 © Shutterstock; 1 © txking / Shutterstock;
22 © fotostory/Shutterstock; 4, 24, 25 © Defenseimagery.mil; 7 top ©
Neveshkin Nikolay / Shutterstock; 11 © Brailleliga, 15, © public domain;
17 top © Featureflash/Shutterstock; 20 © State Farm

Library of Congress Cataloging-in-Publication Data

Wood, Alix.
  Animal handler / by Alix Wood.
      pages cm. — (The world's coolest jobs)
  Includes index.
  ISBN 978-1-4777-6007-9 (library) — ISBN 978-1-4777-6008-6 (pbk.) —
  ISBN 978-1-4777-6010-9 (6-pack)
  1. Animal handling—Vocational guidance—Juvenile literature. 2. Animal
specialists—Juvenile literature. 3. Animal training—Vocational guidance—Juvenile
literature. 4. Animal trainers—Juvenile literature. I. Title.
  SF88.W66 2014
  636.088'8023—dc23
                            2013021149

Manufactured in the United States of America

CPSIA Compliance Information: Batch #W14PK2: For Further Information contact Rosen Publishing, New York, New York at 1-800-237-9932

Contents

What Is an Animal Handler?

Animal handlers work with animals. They work in zoos or animal parks caring for animals or work training animals to perform tasks. They can work for the police, the military, in the entertainment industry, on farms, or helping people with disabilities.

Usually an animal trainer and an animal handler do slightly different jobs. An animal trainer works with animals to teach them desired behavior. An animal handler then works with those animals once they are trained. Working dogs, such as military sniffer dogs, are not usually trained by their handler. They are trained first and then paired with a handler once the dog is ready to start work. Sometimes the same person trains and handles the animal. A **falconer** would usually train and handle his bird himself.

A military police animal handler and her trained dog at work.

FACT FILE

There are different ways of training an animal depending on what type of animal it is. Animals respond to different methods. Handlers need to consider the natural behavior of the animal, and then reward good behavior and take away any reward for bad behavior. Animals need to be trained differently depending on what they will be used for. A hearing dog for the deaf will need different skills from a dog being trained to sniff out explosives.

Hearing dogs need to be trained to alert their owner if a smoke alarm sounds.

👍 THAT'S COOL

Animal handlers often adopt the animal they have been partnered with once it has been retired. Police dogs and military dogs will almost always stay with their handler for the rest of their lives. They build a bond with each other during their working lives.

Even birds can be trained to work for a living. A falcon (below) can be used for pest control. Their natural instinct is to kill mice and other vermin, so it is relatively easy to train them.

Animal Behavior

Any action an animal does or any **response** to a **stimulus** is a behavior. Animals behave in certain ways for four reasons, to find food and water, to avoid predators, to reproduce, and to fit into a social group.

Some animal behavior is instinct, and some is learned. Teaching certain behavior can be useful in helping care for the animal. For example, handlers will teach wild animals to present parts of their bodies for health checks. Animal parks may train intelligent animals such as dolphins to perform tricks to stop them from being bored. To teach a dolphin to jump out of the water, a trainer first rewards the dolphin for touching a target. The trainer uses a long pole with a foam ball on the end as the target. The target is gradually moved higher in the air so the dolphin will jump right out of the water.

Dolphins enjoy jumping out of the water. You can only teach animals to do something that they are naturally happy to do.

FACT FILE

Ivan Pavlov was a Russian **psychologist**. He was one of the first people to discover how you could **condition** a dog to respond to sounds. He carried out experiments with his dog and a bell. Whenever Pavlov fed his dog, his dog would begin to **salivate**. Pavlov began ringing a bell each time he fed his dog. After a while, if he rang the bell, but did not give the dog any food, the dog would still salivate. The dog had begun to associate the ringing of the bell with food, even when no food was nearby.

Ivan Pavlov

👍 THAT'S COOL

Conditioning is important to animal trainers. It's hard to toss a fish to a dolphin while it's in midair. Trainers teach animals to associate something easier to give, like a click on a clicker, with something the animal wants. Once trained, the animal is conditioned to link a click with a treat.

One click on the clicker and the dog knows it will get a treat.

Developing a Bond

An animal handler usually develops a strong bond with her animal. The relationship is similar to how close an owner can feel to his pet, but their working relationship also adds a sense of trust on both sides.

One of the first things a handler will do is try to develop a bond with the animal she is working with. This is usually achieved by feeding and petting the animal and gradually building up its trust. The bond can become very strong.

It's not unusual to hear of police dogs mourning at their partner's funeral, or an army officer becoming depressed if his dog is killed in action. Working with people can sometimes lead to arguments and tensions, but a dog just wants to please its handler. It's easy to see why such a close bond develops.

A working relationship is built on trust.

Building a bond with a dolphin is done with play sessions and food.

FACT FILE

Service animals are trained to perform tasks or help people with disabilities. The bond between a service animal and its handler is very important to the success of the team. Ponies have a longer life span than dogs. They are becoming popular as service animals as the bond made between them can therefore last longer than with a dog. Horses generally have excellent vision. With eyes on the sides of their heads, horses can see nearly 350 degrees around them. They can often detect a potential hazard before their sighted trainers. Horses also have excellent night vision and can see clearly in almost total darkness.

This guide horse at an airport is wearing sneakers! Horseshoes would set off the metal detectors.

Training Service Puppies

Service dog puppies usually live with a family to help **socialize** them before they are old enough to start training. They will learn simple commands and get house trained.

In the United States, a program uses selected prison inmates to train service dogs! It may not seem like it, but it is an ideal environment for the puppies. They spend all their time with their handler, going to work with them in the day and sleeping in their cell at night. The inmates teach the pups around eighty commands. The puppies play with other puppies in the program for at least one hour per day. Puppy raisers will swap dogs so that each puppy gets used to different places and people. **Volunteers** host puppies in their homes at least one weekend per month and take the puppies out on day trips into the community.

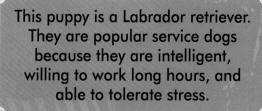

This puppy is a Labrador retriever. They are popular service dogs because they are intelligent, willing to work long hours, and able to tolerate stress.

A guide dog needs to walk ahead of its handler, not "to heel" as most dogs learn. They learn how to do this gradually, by leash commands and spoken commands. They are gradually taught to cope with and ignore distractions. Eventually even the most tempting distraction shouldn't make the dog go off course.

FACT FILE

When a guide dog puppy is about a year old, it starts the next part of its training. It needs to learn

- to walk in a straight line in the center of the sidewalk (unless there is an obstacle)
- not to turn corners unless told to do so
- to stop at curbs and wait for the command to cross the street, or turn left or right
- to never go through a space that is too low or narrow for its handler
- to stop at stairs
- how to deal with traffic

The puppy on the left is being walked through a special training area. The area is set up with different hazards the dog may encounter during its working day. Many training schools have **simulated** streets outside too, so they can expose the dogs to a number of different traffic situations.

Hunting and Herding

Farmers and hunters use animals to help them in their work. They train and handle their animals so that they can work together as a team.

Some animal handlers use **birds of prey** to hunt for them. Handlers start training in a dark room, which makes the bird calmer. The bird gets used to perching on a gloved hand and being fed small pieces of meat. When they start to train outside, the bird is tied to a long cord.

This eagle owl can hunt rabbits.

The handler teaches the bird to hunt for him using a lure. A lure is a piece of leather, feathers, and meat on a long cord. The handler puts the lure down next to the bird and swings it away. The bird learns to catch the lure and bring it back to the handler.

Labrador retrievers are used as gun dogs. They fetch shot birds or other prey and return them to the hunter undamaged. One of the most important skills their handler will teach them is called marking. They must look for falling birds and remember where each bird landed. They are trained to follow the direction of the gun barrel to mark where the birds fall. When the handler commands the dog to retrieve the game, the dog must remember multiple "marks" at once.

👍 THAT'S COOL

Working dogs have a great way of signaling whether they're serious or not. If their tails are up, they're playing. If their tails are down, they're thinking.

Training a herding dog takes patience. A young dog will want to chase the livestock at first. Putting the dog on a long lead and firmly saying "no" should stop this. The easiest commands to teach are "lie down," "that'll do" (come back), and "stand." To teach the dog "come bye" (go left) and "away" (go right), a handler points in the direction he wants the dog to go while saying the command until the dog understands him.

This dog is taking part in a competition called a sheepdog trial.

Helping Dogs Help People

Dogs can be trained to help people in a variety of ways. Animal handlers train these special animals to learn skills that may one day help their new owners live better lives.

A hearing dog helps people who are deaf or hard of hearing by alerting them to important sounds. Doorbells, smoke alarms, ringing telephones, or alarm clocks would all go unnoticed by a deaf person. Outside the home, they can alert their owners to sirens or simply someone calling their owners' name.

Handlers first train the dogs in basic obedience and get them used to things they will meet in daily life. Training usually takes between three months and a year. The dog is taught to recognize a sound and then alert or lead its handler to the source. It may also lead away from a sound, such as in the case of a fire alarm.

Hearing dogs alert their owners to danger by touching them with a paw, then sitting or lying down.

FACT FILE

Some dogs are trained by animal handlers to bark to alert a family when their child has a **seizure.** The dogs also lie next to the children during a seizure to prevent them from injuring themselves. Dogs are thought to detect seizures by noticing small changes in the **epileptic** people's scents or by recognizing tiny muscle movements that happen before a seizure.

👍 THAT'S COOL

The Dickin Medal is awarded for animal bravery. In 2002, guide dogs Salty and Roselle were awarded the medal for loyally leading their blind owners down more than 70 floors of the World Trade Center to safety following the terrorist attack in 2001.

Therapy pets are used to help cheer people up in hospitals. The pets have to be sociable and friendly without being too bouncy. They need to be calm when being stroked. They must take treats gently too, so they don't break a frail person's skin. Handlers test the pets to see how they react to sudden noises – such as a cane falling to the floor. The handlers who volunteer to visit the hospitals with the pets need to be good with people too.

A therapy dog on a hospital visit

15

Animal Movie Stars

Movie animal trainers use their knowledge of animal behavior to train and care for a variety of animals.

Keeping their animals healthy and happy during the hours they are not working on set is important too. Handlers need to sort out travel arrangements, which may involve pet passports, permits, and **quarantine** periods when animals go abroad. Popular movie animals include big cats, reptiles, dogs, cats, horses, bears, elephants, parrots, farm animals, and birds of prey.

👍 THAT'S COOL

The PATSY Award (Picture Animal Top Star of the Year) is a Hollywood honor for animal performers. The award started after a horse was killed in an accident during the filming of the movie *Jesse James*. Animal trainer Frank Inn's animals have won over 40 PATSY awards!

Training an animal for the theater is a bit different from training one for the movies or television. Both have a lot of distractions on the set, but in a theater the animal actor also has a live audience.

Animal stars need to be happy in the spotlight!

It takes time to train a horse to do things that are unnatural for an animal, such as going into small spaces or having someone sit on its back. Unlike dogs, horses are not motivated that strongly by treats. Handlers use methods such as releasing pressure on a harness as a reward for correct behavior instead. Punishment is not very effective either. Horses have a remarkably long memory though, and once a task is learned, it sticks.

Tricks of the Trade

Animal handlers have some cool tricks they use to get the right results from their animals. Most tricks involve understanding what an animal likes or wants and using that to trigger the right response.

Some trainers use a stick with a piece of cheese on one end to help the dog look in the desired direction. Trainers will make unusual noises to get a reaction from an animal too. They may use a phone ringtone or squeaky toys. Sometimes they might make a sound like a kitten or do a loud roar, depending on the reaction they want from the animal.

To make a dog look really happy to see its owner, which could be an actor the dog has just met, the handler may cover the actor's face in bacon oil, which the dog will love to lick off.

This strange group of tools can be used by an animal handler.

Animals in movies or on TV have doubles! A double is another animal that looks identical to the main acting animal. Cats are usually used in teams of identical cats. Using doubles ensures the animals get plenty of rest by sharing scenes, and trained animals can be stunt doubles for an established animal star.

FACT FILE

Movie animals need to be trained to be comfortable on a film set. A handler will often take animals that they are training along with them to a film set while another of their animals is working. This gets the youngster used to the atmosphere and used to spending time away from its handler and mixing with people on the set.

Handling Detection Dogs

Dogs have an excellent sense of smell. Handlers train them using this skill to help us. Dogs can be trained to sniff out items such as drugs or illegal money. They can sniff out explosives too and even help solve mystery fires.

Arson dogs are trained to sniff out traces of **accelerants**, such as gasoline or lighter fluid, that may have been used to start a fire. Each dog works and lives with its handler, who is a law enforcement officer or firefighter trained to investigate fire scenes. The dog learns to recognize different accelerants and is rewarded by its handler for finding one. At a fire scene the dog will search the wreckage until it smells an area where accelerant was used. It will then stop and stare at that point as a signal to its handler.

This dog is training to be an arson dog. It is sniffing cans searching for fire accelerants in the burned-out contents.

When a detection dog finds something, it freezes and stares at it. It'll put its nose on the spot and stay there until it is rewarded. It doesn't bark, growl, scratch, or bite.

Training a search dog is treated like an exciting game. The dog is excited while it is searching because it wants its ball and some affection as the reward. Trainers will scent toys with target odors that they want the dog to search for, and teach the dogs to find the toys. The dogs train in different environments, such as fields, forests, or airports. After six weeks of training, a dog is paired with its handler, a law enforcement officer who will work with the dog. The team then do another six weeks of training together and must pass tests before they can go to work.

An airport detection dog checking luggage.

👍 THAT'S COOL

What if something that was illegal suddenly becomes legal? To retrain a dog, the trainer simply stops rewarding the dog for alerting its handler to that scent.

Search and Rescue

A search and rescue dog's talent in tracking and finding people who are lost or trapped comes from its natural ability, its training, and its close bond with its handler.

A search and rescue dog's training usually begins when it is eight to ten weeks old. Basic obedience is taught using hand signals, as spoken signals may not always be heard in a rescue situation. Agility training teaches the dog to get used to unstable ground, jump through windows, or balance itself while walking along beams.

Searching, tracking, and retrieval skills are taught. Indicating a find to a handler could mean the difference between life and death for a person in need.

After an earthquake in Turkey rescue teams used dogs to help search for survivors.

👍 **THAT'S COOL**

Search and rescue dogs
include water search
dogs and avalanche dogs.
Almost any breed can
be a search and rescue
dog. Larger animals are
preferred due to their
stamina and agility.

To train an avalanche dog,
handlers bury themselves
in snow! This gets the dog
used to searching below
the surface.

FACT FILE

Handlers and search and
rescue dogs need to be so
close they can almost read
each other's minds. The
handler must become aware
of minor changes in the
dog's body language and
notice any small changes
in behavior. Different dogs
may have different responses
when they find a person.
It is up to the handlers
to learn their dog's cues
and respond to them.

Often searches take place on unstable
ground. A dog will do less harm to
people trapped underneath than a
much heavier person would.

Military Dogs

Military dogs are used for guarding military bases, locating land mines and other explosives, and searching for casualties. A military dog handler needs to spend a few years as a regular military recruit, and then apply to become a handler.

As a military dog handler you can specialize in handling one type of military working dog (or MWD). Patrol dogs, drug and explosives detection dogs, and vehicle search dogs are specialities you can work in. Generally the handler will already be a trained member of that unit before being assigned a dog. The military generally run a one dog to one handler policy, so the bond between them becomes strong.

An army dog and his handler search a building.

Military dogs are not treated as pets and are not pets. Unlike many other types of working dogs, they live in a kennel rather than with their handlers. The dogs are taken out many times a day to work and some play time is allowed, too.

It is believed that these living arrangements create a better military dog. The dog's main focus and excitement come from doing its work. The dog is more likely to pay attention and want to please its handler. When it comes out of its kennel, it is happy and ready to work.

FACT FILE

MWDs in the Middle East often have to deal with extreme temperatures, dust storms, and hot sands and **asphalt.** Protective cooling vests, goggles, and boots can help! The goggles protect the animals' eyes from blowing sand. The boots protect paws from hot sand and asphalt, and ear muffs protect the ears from loud noises such as helicopters during air lifts.

Animals in Sport

Animal welfare is very important in animal sports. Trainers must decide if the animals they are training are suitable and enjoy taking part. The most common animals that take part in sports are horses and dogs.

To train a sled dog, a handler gets the dog used to a harness. Then the handler ties objects to a lead attached to the harness, for the dog to drag along behind. Gradually the dog drags heavier objects until it is happy pulling something like a small tire. The dog is then ready to try pulling a sled.

THAT'S COOL

Here are some sled dog commands:
Gee - turn right
Haw - turn left
Whoa - stop
On-by - go past a distraction

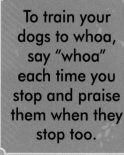

To train your dogs to whoa, say "whoa" each time you stop and praise them when they stop too.

Homing pigeons race to find their way home. To train a pigeon to do this, trainers build a pigeon house with an opening large enough for the bird to fly through. They shut the pigeon inside the house for about 4 weeks. Then they take the pigeon out and gently push it through the opening until it can go in by itself and recognizes the house as home. Now the pigeon should return when it flies away. Once it knows its neighborhood, the bird is taken 1 mile (1.6 km) away and released. Gradually the bird can be released further away in all directions.

These are homing pigeons. Ordinary pigeons do not have a homing instinct.

Show jumpers need a lot of training. To start training a horse to jump, trainers first get it to walk over poles on the ground. Then gradually they make the jumps higher. They always praise the horse after a good jump and end a session on a success.

A horse will generally let you know if it doesn't want to do something!

Still Want to Be an Animal Handler?

If you are interested in a career as an animal handler, start to learn some of the skills you may need. While a marine-mammal trainer may need a degree in **marine biology**, most animal trainers do not have college degrees. A love of animals and lots of patience is more important.

Animals have minds of their own. They behave however they want, until they meet up with an animal handler. Being an animal handler is a rewarding job. The work can be monotonous and sometimes unpleasant, though. Cleaning dog cages every day can be a chore even for the biggest animal lover. Dog bites, horse kicks, and scratches are common injuries for people working with animals.

Handlers are responsible for the day-to-day care of their animals.

While animal trainers work directly with animals, they can have a lot of contact with people too. Dealing with both requires compassion and patience. Teaching animals requires good problem-solving skills. Trainers also need good physical stamina and to be able to bend, lift, and kneel.

People skills are important. You may need to train the owners as much as the animals.

FACT FILE

On a typical day an animal handler will:
- Study her animals to decide what training is needed
- Learn her animals' moods, abilities, and likes and dislikes
- Play with the animals to get them used to a human voice and human contact
- Condition the animals to respond to commands
- Give lots of **reinforcement** when they are good
- Give the animals care, things to do, and exercise
- Oversee the animals' diet preparation

 THAT'S COOL

One of the best ways to get experience working with animals is to volunteer at an animal shelter or other animal-related charity organization. This will give you hands-on experience and a good reference when you apply for a job.

Glossary

accelerants (ik-SEH-luh-rentz)
Substances used to start a fire.

arson (AR-sun)
The illegal burning of a
building or other property.

asphalt (AHS-folt)
A road covering.

birds of prey
(BURDZ UV PRAY)
Birds that hunt and feed
on animals.

condition (kun-DIH-shun)
To train someone or something
to behave in a certain way or to
accept certain circumstances.

epileptic (eh-PUH-lep-tik)
Someone suffering
from epilepsy, which
causes convulsions.

falconer (FAL-ku-ner)
One who hunts with hawks or
trains hawks for hunting.

marine biology
(muh-REEN by-ah-LUH-jih)
A branch of marine science
involving the study of animals
and plants that live in the
ocean and the shoreline.

olfactory (ol-FAK-tuh-ree)
Relating to, or concerned with
the sense of smell.

psychologist
(sy-KAH-luh-jist)
A specialist in psychology.

quarantine
(KWOR-un-teen)
A period during which an
animal suspected of carrying
disease is kept contained.

reinforcement
(ree-in-FORS-ment)
Stimulating (a person or animal) with a reinforcer such as praise.

response (rih-SPONS)
A reaction of a living thing to a stimulus.

salivate (suh-ly-VAYT)
To produce or secrete saliva especially in large amounts.

seizure (SEE-zher)
A sudden attack of abnormal brain activity, with physical signs such as muscle twitching.

simulated (SIM-yuh-layt-ed)
Made to look genuine.

socialize (SOH-shuh-lyz)
To make fit for a social environment.

stimulus (STEM-yu-lus)
Something such as heat, light, or sound that acts to partly change bodily activity, such as by exciting a sensory organ.

volunteers (vah-lun-TEERZ)
People who work freely for no payment helping others or the environment.

WEBSITES

Due to the changing nature of Internet links, PowerKids Press has developed an online list of websites related to the subject of this book. This site is updated regularly. Please use this link to access the list:

www.powerkidslinks.com/wcj/animal/

Read More

Albright, Rosie. *Police Dogs*. Animal Detectives. New York: PowerKids Press, 2012.

Hoffman, Mary. *Helping Dogs*. Working Dogs. New York: Gareth Stevens Library, 2011.

Ruffin, Frances E. *Military Dogs*. Dog Heroes. New York: Bearport Publishing, 2006.

Index